SECRET

> MORALE OPERATIONS
> FIELD MANUAL –
>
> STRATEGIC SERVICES
> (Provisional)

Prepared under direction of
The Director of Strategic Services

SECRET

SECRET

MORALE OPERATIONS FIELD MANUAL

– STRATEGIC SERVICES
(Provisional)

Strategic Services Field Manual No. 2.

SECRET

SECRET

Office of Strategic Services

Washington, D. C.

26 January 1943

This Morale Operations Field Manual, Strategic Services, (Provisional), is published for the information and guidance of all concerned and will be used as the basic doctrine for Strategic Services training for such subjects.

The contents of this Manual should be carefully controlled and should not be allowed to come into unauthorized hands. The Manual is intended for use of the OSS bases and should not be carried to advance bases.

AR 308—5, pertaining to handling of secret documents, will be complied with in the handling of this Manual.

William J. Donovan

Director

TABLE OF CONTENTS

SECTION I — GENERAL PROVISIONS

1. PURPOSE 1
2. DEFINITION 1
3. OBJECTIVES 1
4. IMPLEMENTS 2
5. GENERAL PRINCIPLES 3

SECTION II — ORGANIZATION

6. GENERAL 4
7. MO BRANCH, WASHINGTON 5
8. FIELD ORGANIZATION 5

SECTION III — SELECTION AND TRAINING

9. RESPONSIBILITIES 7
10. MO BRANCH OF OSS, WASHINGTON . . 7
11. FIELD PERSONNEL 7
12. TRAINING OF FIELD PERSONNEL . . . 8

SECTION IV — COORDINATION OF MO FIELD ACTIVITIES WITH OTHER SERVICES

13. COLLABORATION IN DETAILED PLANNING 8
14. COLLABORATION IN OPERATIONS . . . 9
15. RELATIONSHIPS AND LIAISON 9
16. COMMUNICATIONS 10

SECTION V — IMPLEMENTS

17. TYPES 10
18. GENERAL CHANNELS 10
19. PERSONAL CONTACT 13
20. RUMORS 14
21. FALSE LEAFLETS, PAMPHLETS, AND GRAPHICS 18
22. FREEDOM STATIONS 22
23. FORGERY 26
24. BRIBERY AND BLACKMAIL 29
25. COORDINATION OF MO CHANNELS AND IMPLEMENTS 32

SECTION V — PLANNING

26. GENERAL 33
27. BASIC PRINCIPLES OF PLANNING . . . 36
28. OVER-ALL MO PLANS 37
29. TACTICAL PRINCIPLES 37
30. OPERATIONAL PLANS 39

SECTION VII — OPERATIONS

31. GENERAL 40
32. WITHIN THE ENEMY'S COUNTRY . . . 41
33. WITHIN ENEMY-OCCUPIED OR
    CONTROLLED TERRITORY 47
34. WITHIN OTHER AREAS 53
35. WITHIN THE ENEMY ARMED FORCES,
    IN ALL AREAS 55
36. WITHIN OSS, WASHINGTON AND
    THEATERS 58

SECRET

MORALE OPERATIONS FIELD MANUAL
STRATEGIC SERVICES
(Provisional)

SECTION I—GENERAL PROVISIONS

1. *PURPOSE*

The purpose of this Manual is:

a. To present the details of Morale Operations, and their relation to other Strategic Services activities and to OSS objectives, as well as to actual or planned military operations.

b. To set forth the general doctrine to be used in the planning and conduct of morale operations.

2. *DEFINITION*

The term **MORALE OPERATIONS** as considered in this Manual includes all measures of subversion other than physical used to create confusion and division, and to undermine the morale and the political unity of the enemy through any means operating within or purporting to operate within enemy countries and enemy occupied or controlled countries, and from bases within other areas, including neutral areas, where action or counteraction may be effective against the enemy.

3. *OBJECTIVES*

The objectives of subversive morale operations are:

a. WITHIN THE ENEMY'S COUNTRY — To incite and spread dissension, confusion, and disorder; to promote subversive activities against his government by encouraging underground groups, and to depress the morale of his people.

b. To discredit collaborationists, to encourage and assist in the promotion of resistance and revolt against Axis control by the people of these territories, and to raise their morale and will to resist.

c. WITHIN OTHER AREAS BEST SUITED TO THE PURPOSE— To conduct activities which will assist in the accomplishment of the objectives in a. and b. above.

SECRET

d. WITHIN THE ENEMY ARMED FORCES, IN ALL AREAS—To induce low morale and encourage rebellion.

e. WITHIN OSS AND THEATERS—To support and assist other OSS activities, particularly SO and SI, in their respective fields.

4. *IMPLEMENTS*

a. AUTHORIZED*

The Morale Operations Branch, in cooperation with other agencies of OSS, will employ the following implements for the accomplishment of the above objectives:

(1) *Contacts* with and manipulation of individuals and underground groups;

(2) *Agents provocateurs;*

(3) *Bribery and blackmail;*

(4) *Rumors;*

(5) *Forgery*, to include the writing of poison-pen letters, forging of misleading intelligence documents, falsification of enemy documents and periodicals, and the printing of false orders to the enemy, regulations, and proclamations;

(6) *False leaflets, pamphlets, and graphics*, to be used for subversive deception within enemy and enemy-occupied countries and not identifiable with any official or semi-official United Nations agency;

(7) *"Freedom stations"* masquerading as the voice of groups resistant within enemy and enemy-occupied countries when used for subversive deception and not identifiable with any official or semi-official United Nations agency.

b. UNAUTHORIZED

The following implements are not authorized:

(1) *Propaganda and publicity*, including the dissemination of information, arguments, appeals, and

---

* Under JCS 155/11/D, MO activities should confine themselves to means operating within or purporting to operate within enemy or enemy-occupied territory, and should avoid borderline activities such as leaflets which echo official United Nations propaganda themes, even when these leaflets are unlabelled and might be attributed to subversive groups in the target country.

SECRET

instructions by mass means of communications (e.g., radio, press, graphics, motion pictures, official pronouncements) on behalf of or clearly emanating from any official or semi-official United Nations agency. This function is delegated to the Office of War Information.

(2) *Economic pressures* within the jurisdiction of the Foreign Economic Administration.

(3) *Official diplomatic pressures,* which are the formal and informal influences exerted through the medium of our regularly constituted State Department. MO may, however, give unorthodox support to State Department policies and activities when such support is desired and requested by that Department.

5. *GENERAL PRINCIPLES*

a. Focusing on Principal Leaders and Limited Groups

(1) Since the implements employed by MO are not mass means of communication, its operations are focused on principal leaders and special groups who exert, or are capable of exerting, real influence on and control of larger masses of people. These leaders are key enemy military and naval personnel, administrators, civil leaders, quislings, diplomats, and potential leaders of resistance. The groups are special elite groups, important segments of the armed forces, such as, say, naval personnel, soldiers in isolated areas, or soldiers with low morale in certain combat sectors, collaborationists, partisan groups, special political, religious, labor or other organizations, and any disaffected group in the population whose grievances and vulnerabilities can be exploited.

(2) The object is, therefore, to concentrate operations on those critical persons and groups whose subversion or cooperation will produce the most telling effects on large masses. Thus, all morale operations must be carefully "tailored" for the particular persons or groups that are its special targets. Open propaganda methods of wide dissemination and frequent repetition of such appeals as countering enemy

SECRET

propaganda or building up general good-will for the United States are not within the scope of MO strategic services.

b. COORDINATION WITH MILITARY OPERATIONS

(1) MO engages in operations that further actual or planned military operations. In areas of an active military offensive, MO focuses its operations on key enemy military and collaborationist personnel and on specific groups of people at specific times when and at places where its work will be of greatest assistance to our forces.

(2) Similarly, in enemy areas where military action is likely to occur, especially in rear areas of enemy command, communication, supply and transport, MO directs its activities toward critical military or civilian leaders and other persons in specific places where disruption of enemy activity or support of our own projected military action will have the greatest effect.

c. COLLABORATION WITH UNDERGROUND NETWORKS

In order to assist in the promotion of resistance and revolt among people of enemy-occupied and controlled territory and to discredit collaborationists, MO will utilize, so far as is practicable, existing underground networks for MO activities.

d. COOPERATION WITH OTHER BRANCHES OF OSS AND WITH ANALOGOUS ALLIED ACTIVITY

MO works in close liaison with SO and SI and allied networks of operatives and agents, and cooperates whenever possible in joint action. MO tasks in particular areas may be carried out by specially trained MO operatives and agents or by other OSS personnel under MO direction, as the situation requires.

SECTION II—ORGANIZATION

6. *GENERAL*

The organizations within OSS, Washington and in the field, charged with the conduct of morale operations are the following:

a. MO Branch, OSS, Washington, D. C.;

SECRET

<u>b</u>. MO Section, a part of the Strategic Services field base within a theater;

<u>c</u>. MO Section, a part of the Strategic Services base within a neutral area;

<u>d</u>. Field operating personnel, both within theaters and neutral areas.

7. *MO BRANCH, WASHINGTON*

The Morale Operations Branch, Washington, a part of Strategic Services Operations, is responsible for MO planning, administration, and special MO indoctrination of personnel, and for recruitment and procurement, all through established OSS channels.

8. *FIELD ORGANIZATION*

<u>a</u>. ORGANIZATION WITHIN THEATERS

(1) In each theater of operations MO activities are directed by the MO Section Officer under the control of the Strategic Services Officer, or any general Operations Officer designated by the Strategic Services Officer. In some theaters, the theater commander may designate a special unit for deceptive propaganda and combat subversive operations, and call on OSS to assign to it MO personnel, supplies, and equipment.

(2) The MO Section Officer is responsible to the Strategic Services Officer for the procurement through established channels of all MO supplies and equipment in his theater, the training and direction of all MO personnel recruited in the theater, and the administration of all MO personnel dispatched to the field from Washington but not destined for operations in enemy or enemy-occupied territory. He is responsible to the Strategic Services Officer for the direction of and policies and plans governing the operations of MO operatives and agents operating in enemy or enemy-occupied areas.

(3) The following are suggested as assistants to the MO Section Officer:

(a) Staff Officers

The Staff Officers are located in the field headquarters and may include the following persons:

(1) The Deputy Section Officer is assistant to the Section Officer and assumes charge in the absence of his superior.

(2) Liaison Officers maintain active contacts with other branches of OSS, and with other U.S. and Allied agencies conducting analogous activities.

(3) Desk Officers are responsible for plans and operations within their respective designated regions.

(4) The Personnel Officer and such assistants as he may require to recruit, process, and train agents, operatives, and technicians within the theater.

(b) Technicians and Experts

These will include persons with special training in the use of MO technical instruments as required.

(c) Field Operatives and Agents

Field Operatives and Agents will be procured and utilized for field operations, both at the base and behind enemy lines.

b. ORGANIZATION IN NEUTRAL COUNTRIES

(1) In neutral countries, MO personnel will be under the supervision of an MO Section Chief, who will be under the control of the Chief of OSS Mission.

(2) For reasons of security and to avoid interfering with the local requirements of the State Department, the strength of MO personnel should be kept to the smallest number necessary for the effective execution of approved MO programs.

(3) In view of the critical importance of cover in neutral countries, the cover of each authorized person conducting morale operations should be approved in accordance with OSS procedure for cover.

SECRET

SECTION III—SELECTION AND TRAINING

9. *RESPONSIBILITIES*

The MO Branch is responsible for indicating its requirements for personnel, and, subject to OSS regulations, assists the Services Branches in their selection. It is further responsible for indicating to the Schools and Training Branch the types of special training required for MO personnel.

10. *MO BRANCH OF OSS, WASHINGTON*

Personnel of the MO Branch at Washington should have wide and proven experience in the theory and practice of influencing human beings, possessing, if possible, foreign experience and executive and planning ability. The Chief of the MO Branch arranges for such special training of this staff as is necessary.

11. *FIELD PERSONNEL*

Requirements for field personnel vary, depending upon the duties and functions of the persons in question. In all instances due regard for security will be observed. The following is an indication of the requirements of each type of MO personnel:

a. FIELD HEADQUARTERS STAFF OFFICERS

These men should be reliable persons, of United States citizenship, having demonstrated proficiency in administrative affairs and the theory and practice of influencing human beings. They should be familiar with the area in which they are to work but do not necessarily have to have excellent linguistic qualifications. They are not required to enter occupied or enemy territories.

b. EXPERTS AND TECHNICIANS

These persons should be individuals usually of United States citizenship having requisite knowledge and skill to service, maintain, and operate the required MO implements.

c. FIELD OPERATIVES AND AGENTS

These men should be reliable persons, not always

SECRET

of United States citizenship. They must possess linguistic qualifications and a detailed knowledge of the people and area in which they work. Normally they serve in enemy or enemy-occupied territories, but may also be called upon to serve at the base, either within theaters or neutral areas.

12. *TRAINING OF FIELD PERSONNEL*

The training of field personnel both in Washington and in the field includes the regular Strategic Services basic course in secret intelligence and special operations. In addition, this personnel receives special training in morale operations. This training covers:

a. Purposes and organization of MO;

b. Principles of strategic and tactical planning of MO tasks;

c. Methods of clandestine reproduction media, e.g., freedom stations, false printing;

d. Methods of morale subversion;

e. Sources and analysis of current intelligence;

f. Practice on field problems.

Special instruction and training dealing with the area of their activities will be given to all field personnel called upon to operate under cover.

SECTION IV — COORDINATION OF MO FIELD ACTIVITIES WITH OTHER SERVICES

13. *COLLABORATION IN DETAILED PLANNING*

a. Morale operations will be most effective when they are planned as part of common campaigns conducted by various underground services and integrated closely with actual or planned military operations and Allied strategy.

b. The detailed plans prepared by the MO Section Officer should therefore be worked out through the closest liaison with SO, SI, the Allied underground services, representatives of native underground groups,

SECRET

and with military and naval representatives. The strategic services officer facilitates, checks, and approves such coordinated plans.

c. Where specific military operations are contemplated, particular attention is paid to planning MO tasks in support of them. In addition, tasks which facilitate and complement the work of SO and SI are emphasized. The MO planner may be able to assist in the planning of SO and SI missions by drawing attention to the *specific* support that MO can render through bribery, rumor-spreading, use of freedom stations and false pamphlets, forged documents, direct personal contacts.

14. *COLLABORATION IN OPERATIONS*

a. Strategic Services operations out of field headquarters are governed by the nature of the problem that confronts the strategic services officer and by the size and training of the personnel available to him. Morale operations are treated as one integral part of the whole OSS program. When an MO task is to be conducted, and when there are properly placed and trained MO operatives and agents available to do it, these will be used. If a projected task involves physical operations or secret intelligence work, MO men may be given special SO and SI training in order to do the job. Similarly, when an MO task can be best accomplished by SO or SI men without jeopardizing their own work, these men may be given special MO training if necessary and assigned to the job.

b. Whenever SO or SI missions, networks, or contacts are established in a territory, MO should, whenever advisable, operate through them. Similarly, if Allied networks or trustworthy representatives of reliable native underground groups are available and willing to cooperate, these should be employed to do MO work.

c. When specific MO tasks require it or when the theater commander requests it, close collaboration with United States or Allied propaganda agencies in the field should be arranged.

15. *RELATIONSHIPS AND LIAISON*

a. Though MO field activities are under the direction

SECRET

and control of the strategic services officer, the closest relationship between MO Branch, Washington and the MO Field Sections is necessary. Through appropriate channels, the MO Section Officer will report to Washington on his needs, his recommendations, and on the course of his operations. In turn, he will receive from the Director of Strategic Services, Washington, supplies and services which he may require, strategic services programs, plans, and implementation studies, MO campaigns and projects for his consideration, reports on morale operations conducted in other theaters, and reports on morale operations conducted directly out of Washington with which his own work should be coordinated.

b. Liaison with intelligence services at the base in the field is essential to effective MO field operations. This is particularly important for R&A and SI which should be informed as to the needs of MO.

16. *COMMUNICATIONS*

All communications between MO personnel, both in the field and from the field to Washington headquarters, will be through the regularly established communication system.

SECTION V — IMPLEMENTS

17. *TYPES*

The major implements employed in subversive work are: direct personal contact by trained personnel and agents, bribery and blackmail, rumors, false leaflets, pamphlets and graphics, freedom stations, forgeries. The kinds of operations implemented by these instruments are given in Section VII on *OPERATIONS*.

18. *GENERAL CHANNELS*

a. For the purpose of carrying out various morale operations, the MO representative at an Allied or neutral base will make use of specially trained SS agents. However, such agents will be far too few in number to execute all the operations that the MO representative will desire to initiate. It will be necessary to enlist the

SECRET

aid of various individuals and organizations not in OSS employ but strategically placed and willing to cooperate. In many cases, contacts can be made in Allied or neutral territory; in other cases it will be necessary for agents in enemy territory to make the contacts.

b. UNDERGROUND REPRESENTATIVES—Selected personnel of underground groups afford the best means of conducting morale operations, such as discrediting collaborationists, winning hostile elements, and promoting discord and revolt among the population. Hence the closest cooperation is desired. OSS should attempt to comply with reasonable requests from underground leaders for equipment, funds, and moral support, in order that a policy of friendly reciprocity may be established.

c. GOVERNMENTS IN EXILE—Closely related to underground movements are the governments in exile which maintain numerous contacts with the home country; their organizations can be used, where practicable and feasible, for promotion of morale operations. Caution must be exercised, however; certain of the governments in exile have strong political aspirations and may attempt to adapt MO plans to their own purposes. Liaison with such governments in exile should be controlled by the strategic services officer.

d. OTHER U.S. AGENCIES — The State Department, OWI, FEA, the Treasury Department, or the CIAA may be in a strategic position, particularly in neutral countries to provide MO with considerable support in the execution of plans.

e. OTHER ALLIED GOVERNMENTS—In certain areas of the world other members of the United Nations will be in an advantageous position for morale operations, and much time and labor can be saved by collaboration with agencies of these governments.

f. LABOR ORGANIZATIONS—Labor organizations have many connections in enemy territory and in general are among the most vigorous opponents of the enemy regimes. Former labor organizations which have been suppressed by the enemy may still retain remnants of

SECRET

organization that can be used. For example, pre-war trade unions in occupied Europe, though suppressed, may still operate as clandestine organizations. Labor organizations have useful contacts with foreign as well as native laborers in enemy territories.

g. RELIGIOUS ORGANIZATIONS—Religious groups, such as Catholics in Germany, Buddhists in Burma, Protestants in Japan, if properly approached may provide effective assistance for morale operations. Clerics have a high degree of mobility and are less harshly treated by political authorities than ordinary citizens.

h. POLITICAL ORGANIZATIONS—The Communists and the Social Democrats have hitherto been the most active political organizations opposing the enemy within his territory.

i. PROFESSIONAL ORGANIZATIONS — Professional people are, in general, ideologically opposed to the enemy. During the last war, an effective subversive organization in Belgium was composed of professional people (it was run by a university professor and an engineer). Professional classes have numerous international connections and are strategically located.

j. BUSINESS ORGANIZATIONS — Business organizations with branches in enemy territory may still retain connections that can be exploited. This is especially true of business organizations located in neutral countries. The "commercial traveler" is in an especially favorable position for morale operations.

k. FRATERNAL ORGANIZATIONS — Attempts should be made to enlist the cooperation of international fraternal organizations, like the Masons, which may be active in the enemy's territory and opposed to him.

l. NEWSPAPERS AND NEWSPAPER CORRESPONDENTS—Assistance in spreading rumors and making "plants" may be obtained from newspapers and newspaper correspondents, especially in neutral and Allied countries, where stories may be published that will work their way into enemy territory.

m. SAILORS — Sailors from neutral countries whose ships touch enemy ports are particularly useful for

SECRET

spreading rumors and engaging in other subversive work. Contact with sailors in general can most usefully be established through trade organizations.

n. TRAVELERS—Travelers between enemy and neutral or Allied and neutral territories are one of our most common sources of intelligence; it should also be possible to enlist the aid of certain of these travelers in MO activities.

o. FRIENDS AND RELATIVES—The friends and relatives of civilians and soldiers in enemy territory may provide a channel for morale operations in the letters they write to enemy territory.

p. CIVILIANS IN OCCUPIED AREAS — Sympathetic civilians in areas occupied by Allied troops may be allowed to penetrate the lines into enemy territory and carry out morale operations there.

19. *PERSONAL CONTACT*

a. Personal contact is the physical meeting or communication between agents, usually operating under cover, and other persons whose participation in morale operations is desired. The persons contacted are usually individuals in key positions, such as leaders or potential leaders of partisan, dissident, or enemy satellite groups, or any persons whose acts can have important morale-disrupting effects on the enemy.

b. The objective in most instances is to approach persons who can in turn influence others. The purpose is to swing key personnel over to achieving greater cooperation with resistance groups, to swing over fence-sitters and wavering personnel in enemy or satellite areas, to intimidate and terrorize key enemy figures or active collaborationists.

c. The best technique, whenever it can be used, is to employ *persuasion* backed up by promises of supplies. Work of this sort involves appeals to convictions, favors, status pressures, exchange of information, payment of legitimate expenses.

d. A detailed "Who's Who" for the area of operations

SECRET

is essential intelligence for this work. Such a compilation requires a full and detailed list of key native personalities — pro-enemy leaders, pro-Allied leaders, fence-sitters, collaborationists, pseudo-collaborationists, and key figures in organizations in the area. The amount of prestige and influence of each person should be appraised, and a thoroughly worked out description of his background, his strengths and vulnerabilities should be included.

e. Cover, when used in personal contacts, requires the most careful planning. The operative or agent should not only be trained in the principles of cover, and be able to choose one that will protect him from discovery — he should also be "tailored" for the most effective approach to the person to be contacted. He should have the right personality, intelligence, experience, and "contacts" acceptable to the person. The special customs of the area for meeting and negotiating with individuals should not be violated.

20. *RUMORS*

a. A rumor is an unauthenticated, unofficial story or report, represented and transmitted as fact. Two broad types of rumors can be differentiated: subversive rumors which play upon the emotions and attitudes of their audience for the purpose of undermining morale; and rumors which seek primarily to mislead and deceive.

b. Subversive rumors against the enemy are used to exploit the fear and anxiety of those who have begun to lose confidence in military success; to instill false optimism and over-confidence which will lead to disillusionment; to cause popular antagonism to civil and military leaders; to create division among racial, political, religious, and military groups within a country, and between the peoples of allied countries; to cause enemy populations to distrust their own news sources; to lead civilian populations to precipitate financial, food, and other crises by panicky reactions; and to create confusion and dismay by presenting a welter of contradictory reports.

c. In addition to the above, rumors may be used to

SECRET

cause diversionary or impotent enemy action, by revealing false information about our plans and capabilities.

d. The degree to which rumors can achieve such results is determined largely by: (1) the course of military events; (2) the effectiveness of other Allied propaganda media; (3) the effectiveness of enemy propaganda and enemy repressive measures; (4) the emotional state of the audience; and (5) the artfulness and suitability of the rumors in terms of the audiences' background and information. Rarely can they by themselves change basic attitudes. Their function is to confirm suspicions and beliefs already latent; to give sense and direction to fears, resentments, or hopes that have been built up by more materialistic causes; to tip the balance when public opinion is in a precarious state.

e. RUMOR TARGETS

The following groups are most susceptible to rumors:

(1) Groups or classes that have become fearful and anxious about their personal well-being. Rumors should tend to confirm the pessimistic expectations of the group involved.

(2) Groups or classes that have become unrealistically over-confident or hopeful. Rumors should focus on "information" which supports their hopes, which is consistent with information available to them, but which will ultimately produce disillusionment.

(3) Groups or classes that are suspicious of or hate other groups or leaders. Rumors directed against such groups should focus on "information" that justifies and increases hostility.

(4) Groups or classes whose monotonous, humdrum lives make them particularly susceptible to fantasy; for example, men in prisons, army camps.

(5) Highly religious groups.

(6) Primitive, highly superstitious groups whose naive, traditional beliefs can be exploited.

(7) Groups with guilty feelings who fear retribution.

SECRET

(8) Special groups that lack information either as a result of censorship, discredited propaganda, physical isolation, or illiteracy.

f. PROPERTIES OF A GOOD RUMOR

A good rumor is one which will spread widely in a form close to that of the original story. Probably the main factor determining whether it catches on is the degree to which it is adapted for the state of mind of the audience. In addition, successful rumors embody most of the following qualities:

(1) *Plausibility*. A plausible rumor is tied to *some* known facts, yet incapable of total verification. It may exaggerate, but it stops short of the incredible. It frequently appears as an "inside" story.

(2) *Simplicity*. A good rumor uses only one central idea as a core. Its basic message is simple and thus easy to remember.

(3) *Suitability to task*. To summarize opinions or attitudes which are already widely accepted, slogan-type rumors are best. ("England will fight to the last Frenchman"). To introduce "information" which will help build up *new* attitudes, however, narrative-type rumors are best (e.g., rumors which "prove" that Hitler is mentally ill).

(4) *Vividness*. Regardless of length or type, rumors which stimulate clear-cut mental pictures with *strong emotional content* are likely to be most effective.

(5) *Suggestiveness*. The type of rumor which merely hints or suggests something instead of stating it is well adapted to spreading fear and doubt. The listener should always be allowed to formulate his own conclusions.

(6) *Concreteness*. The more concrete and precise a rumor, the less likely it is to become distorted in transmission.

g. USING RUMOR CHANNELS

(1) To do an effective job of dissemination, a thorough and systematic survey must first be made of

SECRET

all possible channels. The form and content of a rumor should then be adapted to the particular outlets through which they are to be launched. For example, a rumor to be spread via longshoremen's circles might have a bit of pornography as its core. The several channels should be regularly re-examined to determine whether maximum circulation is being achieved.

(2) Rumor-mongers should be natives of the territory where the rumors are to be spread. For reasons of security, it is important that the rumor-monger should be discreet. The rumor should be spoken before an innocent but talkative person (e.g., barber, bartender) who will unconsciously spread the desired rumor.

(3) Rumors can also be initiated by planting material in newspapers, by writing provocative letters containing a purposive message, leaving the letters lying about where people will pick them up when they are alone, by use of denials, and by studied indiscretion.

h. COORDINATION AND TIMING

There are no inflexible rules as to whether rumors should precede or follow up other subversive implements in the execution of a given campaign. Rumors frequently prepare the ground for the later use of subversive pamphlets, for personal recruiting, and for open propaganda broadcasts and leaflets. In other instances, as, for example, following SO demolition, rumors may be used most effectively to exploit and exaggerate results achieved. To catch on, rumors must usually exploit the momentary focus of public interest. Hence, one rule of timing is always applicable: there must be as little time-lapse as possible between the provoking event and the launching of the rumor.

i. SPECIAL USES OF RUMORS IN CONNECTION WITH MILITARY OPERATIONS (to be employed only as directed by the theater commander).

The most effective way in which rumors can be used in direct connection with military operations is

SECRET

in the creation of panic and confusion among civilian populations and enemy troops. Fright rumors can be used to direct civilian refugees into the path of a fleeing enemy and thus block his retreat. They can be used against enemy forces to induce surrender. Rumors may also be employed to foster or allay guilt, i.e., by attributing violence or atrocities to enemy forces or to our own, as the situation may require. Following military action by our forces, they can be used to create fear and anxiety by exaggerating the magnitude of damage done. They have also been used effectively following military action to cause the enemy to release, through refutation, information of value to our intelligence services.

j. RUMOR INTELLIGENCE

The basis for good rumors is accurate, detailed intelligence. The rumor planner and the rumor operator must keep the closest possible check at all times on the character of the group they are trying to affect, on their traditions, circumstances, sentiments, and interests, and on contemporary happenings and developments. It is essential to have intelligence on what the audience knows and what it does not know, on what it fears and hopes for, on what its morale is at any given time, on what kinds of rumors have "caught on" in the past in the particular area. In many cases, the most effective rumor policy will be to spread further rumors that have arisen spontaneously in enemy territory.

21. *FALSE LEAFLETS, PAMPHLETS, AND GRAPHICS*

a. This type of implement refers to printed, mimeographed or written literature and graphics distributed secretly in enemy territory and under concealed sponsorship. This includes chain and other anonymous letters, chalking symbols and messages on walls. The false pamphlet sponsored by a belligerent nation attempts to convey the impression that it is a bona fide message from the people's own fellow country-men who are sharing the same risks as the rest of the population and have similar aspirations, aims, and goals.

b. An appeal to nationalistic attitudes is more effective when made by the nationals of the group than when

SECRET

made by another nation which has its own nationalistic axe to grind. Likewise, incitement to action or revolt coming from a representative of an aggrieved group is more effective than such incitement coming from an "outsider." Whenever an attempt is made to assure potential sympathizers that they would not be alone in resistance activities, such assurance comes better from a group which is presumably carrying on the same activities under the same conditions and taking the same risks.

c. USES OF FALSE LEAFLETS

(1) In general, the false leaflet can be used for dissemination of "forbidden" news, spreading of rumors, exposing nefarious activities of enemy officials and collaborationists, giving reassurances to potential sympathizers, instructing in sabotage, inciting to subversive activities, and preparing the populace for cooperation with invading troops. The false leaflet can be capitalized upon by propaganda agencies in popularizing a passive resistance campaign (such as the "V" campaign or the "1918" campaign). It may be desirable to have the campaign "planted" via false pamphlets. This "spontaneous" activity can then be picked up by the authorized propaganda agencies.

(2) The false pamphlet can be effectively used to make the enemy uneasy about the loyalty of the people in the territory. The very existence of clandestine pamphlets is "evidence" of underground activity.

d. OPERATIONAL PROBLEMS AND LIMITING FACTORS

(1) While false literature ostensibly originates in enemy-occupied territory, it is frequently printed in neutral or Allied countries adjacent to the countries for which it is destined. This immediately raises important questions of (a) authenticity, (b) procurement of supplies, (c) transportation and distribution, (d) timing, and (e) security.

(2) The kind of paper, typography, ink, diction, colloquial expressions, and general composition should be indigenous to the country. That necessitates having, in the Allied or neutral country, the

SECRET

proper supplies and equipment for printing the pamphlets, and the personnel to compose the pamphlets; or providing natives within enemy territory with proper supplies, equipment, and instruction in clandestine methods of reproduction.

(3) The problem of distribution is two-fold: first, the pamphlets, cuts, mats, and other supplies often must be smuggled into the enemy-held country by boat, submarine, plane, or agent; and second, the pamphlets must be distributed to the ultimate consumer within that country. The simplest method of distribution — scattering pamphlets from Allied planes — is not feasible since this would immediately betray the real origin of the pamphlets. The airplane, however, may be used to make pre-arranged wholesale deliveries to resistance groups who would then be responsible for retail distribution. In any event, distribution to the final consumer requires a network manned by native personnel in the country.

e. TIMING

In the case of pamphlets that are printed outside the country, the transportation and distribution process will frequently be a slow and dangerous one, and a considerable time will elapse between the period when a pamphlet is printed and the moment it reaches the intended reader. Therefore, such false leaflets cannot be used to exploit events of immediate and temporary importance.

f. SECURITY

(1) The problem of security is an important one *at every point of the process*. The printing, whether done externally or internally, must be kept secret from enemy agents and even from the people to whom the pamphlets are addressed, else the Allied origin would be disclosed. The smuggling of pamphlets or supplies into enemy territory is extremely hazardous and requires careful planning. Finally, the internal distribution of the pamphlets endangers the security of both the distributor and the reader.

(2) As a practical security aid, the pamphlet

should preferably be small, probably not greater than 6 x 8 inches. This facilitates smuggling and distribution, and also tends to protect the recipient of the pamphlet by making it much easier to hide. Not only should the pamphlet be small, but the message itself, whenever possible, should be brief. The message should stress one point and suggest positive action.

(3) It is customary to conceal false pamphlets by using the cover and format of a familiar enemy publication, such as a magazine, newspaper, or time-table; and by inserting them in packaged goods, containers of various sorts, books, magazines, and other pamphlets.

g. INTELLIGENCE

False pamphleteering can be successfully executed only if based on accurate information regarding the current attitudes and morale of the group for which it is adapted. Various other types of expert information will be necessary depending on the nature of the pamphlet. For example, a pamphlet that gives instructions on sabotage must be based upon detailed technical and regional knowledge.

h. COORDINATION WITH OTHER AGENCIES

Several different agencies and types of personnel must cooperate in false pamphleteering. MO's function is primarily that of writing and translating texts, and directing distribution. The distribution of the pamphlets will usually be made by sympathizers and members of underground groups in the country. The information which goes into the pamphlet must be obtained from every source possible, including SI, SO, FN, and R&A. Natives of the country should be recruited by MO for the technical job of composing the pamphlets in order to insure authenticity in idiom and expression. For special MO campaigns, false pamphleteers should coordinate their activities with those of propaganda agencies, within limits of security. False pamphlets that urge sabotage or open revolt must be worked out under coordination with the theater command.

SECRET

22. *FREEDOM STATIONS*

Various types of *cover* are used by freedom stations. They may pose as: the organ of a subversive (freedom) group within an enemy area; a regular official station of the enemy; the special organ of an official enemy group (Army, Navy, or party); the organ of an anonymous or private group within enemy territory, not openly subversive. Some freedom stations have no very definite cover at all beyond that of being a freedom station run by a colorful personality, the question of identity being left as a mystery.

<u>a</u>. Uses of Freedom Stations

Freedom station transmissions can be used for the following purposes:

(1) To spread demoralizing rumors among enemy soldiers and civilians;

(2) To encourage "patriotic" resistant groups within enemy territory by acting as their spokesman;

(3) To stimulate and direct sabotage and subversive activity;

(4) To make preparations for, and direct, popular uprisings and pro-Allied activity on D-day;

(5) To terrorize collaborationists and Axis officials by black-balling, giving the impression of a vast and powerful underground;

(6) To divide the enemy group from group, or nation from nation, by spreading divisive stories, by posing as a representative of one group or nation and condemning, insulting, ridiculing another group or nation;

(7) To heckle enemy broadcasts.

<u>b</u>. Establishing a Freedom Station

(1) In addition to technical experts, the staff of a freedom station includes broadcasters who speak the language of the area to which the transmissions are beamed without a trace of foreign accent and with full knowledge of current slang. Freedom station broadcasters must appear to be typical representatives of the intended audience.

SECRET

(2) Whether short or medium wave is to be used depends upon the distance from transmitter to audience (much greater distance with the same power can be achieved by a short-wave transmitter) and upon the type of receiving sets possessed by the audience (in Europe medium-wave receiving sets are much more common than short-wave). Mountainous obstructions reduce the range of audibility. Short-wave transmission is very good across water, very poor across sand.

(3) Ordinarily a freedom station is permanently located in friendly territory as close to the enemy area as power supply, transportation, and political problems will permit. The audience is gradually built up. On occasion, it may be feasible to employ mobile transmitters or captured enemy transmitters for freedom station purposes.

c. INTELLIGENCE

A freedom station must have detailed, up-to-the-minute intelligence. The best sources are: interviews with travellers from enemy areas, interviews with prisoners of war, reports from agents in the field, censorship material. Excellent intelligence is necessary because the clandestine station pretends to be closely identified with its audience, and to be broadcasting from within enemy or enemy-occupied territory; and because much of the audience appeal of subversive broadcasts depends upon its ability to reveal intimate intelligence about personalities and local affairs. While unverifiable rumors may be spread on occasion, it is important to hook all broadcast materials onto facts which can be verified by the audience, thereby lending prestige to the medium and making the rumors more credible.

d. AUDIENCE TECHNIQUES

Freedom stations employ various techniques to obtain a larger listening audience such as:

(1) Emphasizing the very mystery and danger of being a "Freedom Station," changing frequencies suddenly, or breaking off transmission and returning at a later period;

SECRET

    (2) Providing "inside dope" — naming names, giving facts in detail, giving the impression of eye-witness account;

    (3) Using slang, vulgarity, and pornography when consistent with cover, giving gossip and "dirt," e.g., describing the sexual life of prominent enemy officials or their wives;

    (4) Providing local color through news items, songs;

    (5) Making special group appeals to youth, women, peasants, workers, Catholics, social democrats;

    (6) Selecting a domestic "enemy" or object of hate (individual leader, political, social or ethnic group), continually attacking him and blaming him for all evils;

    (7) Predicting military developments;

    (8) Pretending to have a tremendous following, appearing to address organized groups — "comrades";

    (9) Using catchy titles, symbols, slogans, songs, jokes;

    (10) Providing news "scoops";

    (11) Playing incidental recorded music which will catch "dial twisters";

    (12) Making broadcasts brief (10-15 minutes), so that the danger to listeners will be minimized and the chances of hearing a particular message maximized;

    (13) Broadcasting on enemy domestic frequencies during lulls, between programs, and pauses in speeches during air raids, and shortly after enemy radios shut down for the night, thus catching the regular enemy audience. Since a single freedom station program uses a transmitter only a small part of the time, it is common to use the same transmitter for several different programs, usually directed to different audiences.

e. SECURITY

    (1) The freedom station presents an unusual

SECRET

security problem. Extreme precaution must be taken not to reveal its sponsorship or location. Nevertheless, there is little hope that its *general* sponsorship and location can be concealed from the enemy, who will spot it through radio engineering methods, and therefore it must be carefully concealed to prevent bombing attacks. Furthermore, while enemy officials may discover the general location and sponsorship, the chances are that they will not reveal the information to their own public, because it would only call attention to the freedom station and possibly increase its audience (there is no record of an enemy country ever having "exposed" a freedom station).

(2) An additional security problem arises from the fact that a freedom station is frequently so indecent and deceitful that its exposure might arouse public opinion in one's own country and result in public arguments on the moral issues involved.

(3) Special precautions must be taken that the station does not present information which, in view of its cover, it could not be expected to know; or information which comes from a recognizable source.

f. COORDINATION WITH OTHER OPERATIONS

A freedom station may assist SO operations in the following ways: by giving instructions on, and stimulating sabotage and guerrilla warfare; by acting as spokesman for and encouraging an underground movement; by directing popular uprisings and pro-Allied activities on D-day.

g. COORDINATION WITH MILITARY OPERATIONS

A freedom station can assist the military in a theater of war by spreading confusion, demoralizing, and divisive rumors; by encouraging resistance to the enemy; and by directing sympathetic groups on D-day.

h. LIMITATIONS OF THE FREEDOM STATION

The effectiveness of a freedom station is limited by the following conditions: the power of the transmitter; the number of receiving sets possessed by the audiences; the willingness of individuals in enemy areas to risk their lives listening to a forbidden radio program; the severity of punishment for listening; the trust which listeners will put in a broadcast that professes to speak in their interest but has no authority behind it. The freedom station is also limited by the degree to which the listeners take the station seriously and do not feel they are being hoaxed. A few slips, such as out-of-date slang, misinformation easily checked up, or failure to maintain cover, will quickly show up a clandestine broadcast.

23. *FORGERY*

a. "Forgery" is understood to mean the act of falsely making or altering any kind of document.

b. Forgeries can be used for the following purposes: to assist the theater commander, when requested, to deceive and confuse the enemy regarding our military intentions and capabilities; to harass and over-burden the enemy administration; to implicate enemy personnel; to spread information and rumors designed to demoralize and divide the enemy; to arouse false hopes.

c. Forgeries are usually intended to fall into the hands of enemy military personnel, enemy political police, or groups who may possess definite suspicions that will be confirmed by the forgery, and to implicate enemy personnel and destroy confidence in leaders.

d. Forgeries will usually be the more effective if supported by rumors and other morale operations. Conversely, forgeries are particularly important for confirming operations initiated through other channels.

e. Extreme care must be taken to make the forgery appear *authentic* — otherwise it may have a serious boomerang effect. It must be technically perfect. It must fall into the proper hands through apparently normal channels. It must be consistent with facts known and impressions currently held by the enemy, preferably adding fuel to an already existing fire. Too many forgeries will defeat the purpose — unless the purpose is that of making all documents of the given kind suspect. A poor forgery may not only expose an agent, but may also expose and invalidate other more competent forgeries.

SECRET

f. TYPES OF FORGERIES

Any kind of written communication can be forged. The following are common types: (1) propaganda documents, ostensibly designed to increase the morale of the enemy or improve his relations with an enemy ally, but actually written in such a way that an opposite effect will be achieved; (2) periodicals, which imitate enemy periodicals and convey misleading or morale-disturbing information; (3) business documents, using letterheads or other business forms of either enemy, Allied or neutral firms, and filled out with misleading information; (4) cables, either Allied or enemy; (5) canards, which are extravagant reports or documents circulated for the purpose of deluding large sections of the public; and (6) letters of all kinds. In addition to the above, upon request of the military authorities, the Commander can be assisted by forging military documents, which include fake *enemy* military documents, especially fake orders; and fake Allied military documents, especially documents containing false intelligence on Allied plans and capabilities.

g. INCRIMINATING DOCUMENTS

(1) Incriminating forgeries, especially letters, are prepared to cast suspicion upon selected enemy or collaborationist personnel actually loyal to the enemy. Letters may be sent to the individual in question from a neutral country and be incriminating by virtue of the implications of their contents when intercepted by enemy censors; they may be sent from one neutral country to another and be intercepted when passing through enemy territory; they may be sent to enemy police, party or military officials from a neutral country, and contain direct charges; they may be sent anonymously from within enemy territory to police, party or military officials and contain direct charges; or they may be left in spots where enemy officers will "discover" them. The latter type is dangerous, however, unless expertly done (preferably by pasting together words from newspapers) because police efforts will doubtless be made to discover the writer.

SECRET

(2) Incriminating documents are often designed merely to harass and over-burden the enemy secret police. If supported by other evidence, they may, however, succeed in causing liquidation of undesirable persons although this is a result that should not be anticipated unless circumstances are very favorable and the case is built up with the greatest skill. Additional results of incriminating documents may be resentment against the secret police on the part of individuals investigated (particularly of army officers), and the slackening of secret police investigation of really disloyal personnel, should a number of suspect individuals turn out to be loyal and cause embarrassment to the secret police.

(a) *Targets*. Collaborationists provide good targets because enemy police are more likely to be suspicious of these individuals than of their own people. Certain collaborationists have, in fact, turned out to be really working for the Allies. For this reason extreme care must be taken in selecting the target. One must be absolutely sure that he is in fact a quisling. A thorough and accurate intelligence report must therefore be first obtained on each potential target.

(b) Army or navy officers are usually good targets in view of the internal discord that would be created by a secret police investigation.

(c) Officials of enemy-allied or satellite nations, usually suspect in the eyes of the enemy, make good targets. Only those known definitely to be enemy sympathizers should be concentrated on.

(d) If a forged document is directed against a party official, he should ordinarily not be so high in rank that the lower officials will be afraid to do anything about it, or will refuse to believe it.

(e) *The incriminating case*. The offense charged or implied should be serious enough to be worthy of secret police attention. It should be consistent with the known past of the accused. For example, a man of wealth, particularly one who has often dealt with foreign firms, would be a most

SECRET

suitable person to implicate as having concealed funds in a neutral country. A direct accusation should be accompanied by several correct corroborating circumstances. A mere general charge that X is really not a quisling but an Allied agent is likely to receive little attention, but a statement of true particulars of X's whereabouts and actions on a certain day on which he is alleged to have acted treasonably will produce action.

(f) A letter of accusation should be written in the way in which the purported sender would be expected to write. If, for example, it purports to come from an Argentine banker and to acknowledge to a Swiss banker the receipt of funds for the account of a French client, ordinary commercial language should be used. The accusation should not be so subtly concealed that the censor or other intended reader will fail to discover it. A number of incriminating letters accusing different people should not be sent from the same source in the same style, with the same paper.

24. BRIBERY AND BLACKMAIL

a. Bribery and blackmail, while in many cases extremely effective, must be used with great caution. Unless done skillfully, they may result in exposure of the operator. This is especially true for bribery, since the art of double-crossing is an ancient one, and the bribed individual is apt to be an unscrupulous person willing to work for either side.

b. Bribery and blackmail must be adjusted to the social customs and expectations of the recipient. In some areas of the world (particularly in the East) and among some classes of people, bribery is almost as common as tipping in the United States; in other areas and among other classes of people, the mere suggestion of bribery is highly insulting.

c. Such individuals as political and military leaders, newspaper editors and reporters, radio broadcasters, heads of business houses, religious, professional and

SECRET

labor leaders, police, petty officials, customs officers, and sentries are the most useful targets of bribery.

d. USES OF BRIBERY

(1) In rare instances, bribery may be effective in accomplishing important strategic secret diplomatic acts, especially in enemy satellite countries (see paragraph 33-e). Typically, bribery is used to *aid* in carrying out less ambitious operations. Thus, for the purpose of spreading rumors, it may be desirable, especially in neutral areas, to bribe newspaper men or radio announcers to plant the rumors in newspapers or in broadcasts. Bribing of police officials may facilitate the creation of an "incident" or riot.

(2) In some cases, bribery can, by itself, achieve certain MO objectives. Thus by judicious bribing of local leaders of various enemy or fence-sitting political, religious, labor or professional groups, the group may be induced to engage in subversive work — or by selective bribing of such officials, dissension might be created in the rank of the organization. Successful bribery of an enemy or collaborationist official followed by exposure to the enemy authorities can also be used to discredit or neutralize the effectiveness of such officials and create doubt and suspicion of all officials.

e. OPERATION OF BRIBERY

(1) It is often desirable for the first services purchased to be of a minor character, and one not involving great risk on his part. Once the initial bribe has been accepted, and evidence of such bribery has been obtained, the demands can become successively greater. Where possible, the "reward" or bribe should also be of such a nature as to become increasingly indispensable to the recipient.

(2) In many cases money may be less effective than goods or services, particularly in areas where certain goods and services are relatively inaccessible while money is plentiful. The following may be useful depending on the needs and susceptibilities of the recipient: food, medicines, drugs (this may involve first inducing a dependency in the individual upon a

drug), clothes, liquor, employment, escape to neutral countries, transportation of letters to friends and relatives outside, release of relatives or recipient from prison, protection, business tips, social and political favors, especially aid to the recipient's family.

(3) Indirect or covert bribery may be used where it will reduce the danger of exposure and avoid the possibility of insulting the recipient. Covert bribery involves the use of such techniques as the following: selling goods below their value; buying goods above their value; losing to the recipient at gambling; making unwinnable bets with him; presenting him with expensive gifts; making heavy "loans"; granting monopolistic rights to certain revenues, products or services; establishing "philanthropic" organizations as fronts; subsidizing corporations.

f. TYPE OF INTELLIGENCE REQUIRED

To carry through successful bribery it is essential to have full intelligence on the character of the recipient — his needs, weaknesses, grievances, fears, hopes, honesty, and integrity. What he feels deprived of in the way of goods and services should be known. Closest collaboration with X-2 should be maintained.

g. BLACKMAIL

(1) Blackmail is directed against the same targets and can be used for the same purposes as bribery. It differs from bribery in that threats, rather than rewards, are used to induce action. They are ordinarily threats to divulge information about the individual which would cause him serious harm, socially, politically, or physically. This information can be based on acts committed by the individual in the past, acts now being committed, acts which the individual *believes* he has commited, or acts which he has not in fact committed but for which evidence is planted against him.

(2) Threats to reveal infractions of rules — especially military regulations — constitute a good hold on a man. In wartime, regulations are so numerous, complex, and severe, that it is difficult for any-

SECRET

one not to break *some* of them, or to fear one has broken them.

(3) Blackmail is often a secondary stage of bribery. After a man has accepted bribes, he is kept in line by threats that his duplicity will be exposed; or both may be used at the same time.

(4) The action demanded of the victim should be consonant with the blackmail risk involved. That is, care should be taken to see to it that the action demanded of him does not entail greater risks than the consequences of exposure of his original "misdeed."

h. TYPE OF INTELLIGENCE REQUIRED

(1) Blackmail requires intimate intelligence on the man, his family, his friends, or his associates. Documentary or photographic evidence is valuable.

(2) When incriminating information is difficult to secure, or when no such information exists, it may be possible either to create it or to plant it. A study of the individual's character should suggest special vulnerabilities (drugs, alcohol, women, luxury, power), which may be exploited and result in transgressions. If this is not feasible, spurious documents, and bits of "telltale evidence" can be used to build up a case. The latter procedure, however, will be extremely difficult and dangerous. Closest collaboration with X-2 should be maintained.

i. COORDINATION WITH OTHER OSS AGENCIES

Inasmuch as bribery and blackmail are two techniques which will also be used by SO and SI, projected operations should be checked with these branches. Frequently an individual who has been successfully used by one branch can be used by the other.

25. *COORDINATION OF MO CHANNELS AND IMPLEMENTS*

a. A morale operation will be the more effective if several implements are used to carry it out. For example, a rumor or an anti-Axis slogan may be planted in a freedom station broadcast. A few days later it may be

SECRET

carried also in a false pamphlet. Through personal persuasion or possibly bribery, a newspaper in a neutral capital picks it up. Evidence (forged) is discovered confirming it. Finally the propaganda radio picks up the item.

b. Certain channels naturally supplement each other. The same appeals can be made by freedom station and false pamphlet. In fact, it is common for false pamphlet writers within enemy territory to obtain much of their material from clandestine broadcasts. Bribery and blackmail may be employed to facilitate rumor-spreading and false pamphlet distribution. Forgeries will confirm suspicions about individuals initially raised by freedom stations, false pamphlets, or rumors. And conversely, a false pamphlet, for example, can photostat and publicize an incriminating forged document.

c. When several implements are used to carry out an MO project and when the project is developed through the coordinated use of several implements and channels, the operation is referred to as an MO "campaign". The most important projects should, if possible, be carried out in this fashion.

SECTION VI — PLANNING

26. *GENERAL*

a. NOMENCLATURE

For convenience in planning, the meanings of the following general terms are indicated below:

(1) *Over-All Program for Strategic Services Activities;* a collection of objectives, in order of priority (importance) within a theater or area.

(2) *Objective;* a main or controlling goal for accomplishment within a theater or area by strategic services as set forth in an Over-all Program.

(3) *Special Program for Strategic Services Activities;* a group of detailed missions, assigned to one or more Strategic Services branches, designed to accomplish an objective, having also a general statement of the situation and the general methods of accomplishment.

(4) *Mission;* one or more subsidiary goals set forth in a special program which together result in the accomplishment of an objective, and are usually assigned to one Strategic Services branch.

(5) *Operational Plan;* an amplification or elaboration of a special program, containing in greater detail the means of carrying out the specified activities.

(6) *Task;* a detailed operation, usually planned in the field, directed toward the accomplishment of a mission.

(7) *Target;* an individual place, establishment, person, or individual which is involved in the accomplishment of a task or result desired.

(8) *The Field;* all areas outside of the United States in which Strategic Services activities take place.

(9) *Field Base;* an OSS headquarters in the field, designated by the name of the city in which it is established, e.g. OSS Field Base, Cairo.

(10) *Advanced or Sub-Base;* an additional base established by and responsible to an OSS field base.

(11) *Operative;* an individual employed by and responsible to the OSS and assigned under special programs to field activity.

(12) *Agent;* an individual recruited in the field who is employed and directed by an OSS operative or by a field or sub-base.

(13) *Cover;* an open status, assumed or bona fide, which serves to conceal the secret activities of an operative or agent.

(14) *Cutout;* a person who forms a communicating link between two individuals, for security purposes.

(15) *Resistance Groups;* individuals associated together in enemy-held territory to injure the enemy by any or all means short of military operations, e.g., by sabotage, espionage, non-cooperation.

(16) *Guerrillas;* an organized band of individuals in enemy-held territory, indefinite as to number, which

SECRET

conducts against the enemy irregular operations including those of a military or quasi-military nature.

b. STRATEGIC TASKS

In deciding on the general MO strategy for a given territory, the major subversive tasks most suitable to the situation existing in the area are selected. Below are listed the major tasks which may achieve the MO objectives within the areas cited in Section I, paragraph 3.

(1) *Within the enemy's country*

    (a) Dividing the enemy.

    (b) Inducing panic in enemy population.

    (c) Strengthening the enemy civilians' desire for peace, and raising false hopes.

    (d) Subverting enemy civilian populations during a ground offensive.

(2) *Within enemy occupied or controlled territory*

    (a) Promotion of resistance and revolt against the enemy.

    (b) Interfering with the enemy's consolidation and use of an occupied country's capabilities.

    (c) Producing civilian disorder in support of military operations.

    (d) Increasing terror, friction, and demoralization among collaborationists.

    (e) Provoking rebellion or a *coup d'etat* in a satellite country or inducing its separation from the Axis.

(3) *Within other areas*

    (a) Establishing an MO section of a base in a neutral country for operations in enemy and enemy-dominated countries.

    (b) Assisting the chief of the diplomatic mission in special work requested by him.

(4) *Within the enemy armed forces, in all areas*

    (a) Fostering rebelliousness within enemy armed forces.

    (b) Inducing surrender.

SECRET

(5) *Within OSS, Washington and theaters*

(a) Within OSS, Washington, over-all planning for and servicing of field operations.

(b) Assistance desired by the theater commander.

27. BASIC PRINCIPLES OF PLANNING

All plans prepared by MO must be for the purose of attaining the purpose of the over-all program for a given theater or area and of the special MO programs pertaining thereto. In such inplementation the following general principles should be observed:

a. SUITABLE OBJECTIVES — Successful planning and execution of MO tasks require that *specific* MO objectives be clearly delineated. These objectives are those which, when fully achieved, produce the most damaging morale disrupting effects on the enemy.

b. FEASIBILITY — Only those MO tasks are planned in which all the operations required to accomplish the objectives are practical of execution with the means OSS possesses or can reasonably expect to obtain from the theater commander and other cooperating agencies or media.

c. ACCEPTABILITY — All subversive operations should lie within the scope of MO policy as authorized, and should justify the costs involved. Since MO activities by their nature are deceptive and not attributable to any official or semi-official United States or United Nations agency, morale operations need not conform to public statements of policy. However, they should be so planned and executed as not to jeopardize United States state of war policy.

d. OFFENSIVE — In general, MO tasks should consist of aggressive, determined, continuous, and unrelieved action against the enemy.

e. TIMING AND PLACING—The military, and especially the political and social situation in which the task is to be carried out, should be carefully studied and the subversive action planned for the exact time when, and place where, its total effect will be greatest.

28. OVER-ALL MO PLANS

All subversive operations in a given territory should be in conformity with an objective in a special program. MO should prepare MO plans for the implementation of these special programs and although no rigid form is necessary for such a plan, the MO section of it should, in conformity with the basic principles listed above, contain the following:

a. A brief statement of the missions in the special programs that MO is charged to accomplish.

b. A succinct survey of the situation, including only data that has *direct bearing* on the MO tasks which will carry out the stated missions. This survey would normally include summaries of the military situation, of the capabilities of the enemy, and of the capabilities of MO and its cooperating agencies and media.

c. A listing of the proposed tasks or courses of action which will achieve the stated missions. If necessary, order of priority should be given, and any timing or placing requirements. The cooperative relation of these tasks to military operations, to other strategic services operations, to subversive strategic plans of Allied partners, and (when relevant) to MO plans in other territories and theaters should be shown. It is to be assumed that all the tasks have been checked for acceptability, but if any question is to be anticipated, a statement clearing up the matter should be included.

d. A statement of requirements for the projected tasks, including an indication that arrangements for carrying out all the necessary operations have been made.

e. If necessary, a statement of any important contingencies that may seriously alter the plan, and how these contingencies will be met.

29. TACTICAL PRINCIPLES

a. Tactics are the particular morale operations necessary to carry out the general tasks laid down in the over-all MO basic plan. Operational planning requires

SECRET

decisions on what specific tasks are to be engaged in, precisely when and where the job is to be done, what specific operatives, agents, collaborating persons are to be employed, what particular persons or groups are the objects of subversive action, and what implements are necessary.

b. Originality, knowledge of the situation, enterprise, and practical judgment determine good tactics. Nevertheless, specific courses of subversive action are always more effective if governed by a few simple principles. If these are ignored, results are often ineffective or wasteful, if not disastrous.

c. The principles of strategy apply equally to tactics—those of object, offensive, timing and placing, feasibility, and acceptability. In addition, there are a few principles dealing specifically with subversive tactics.

d. Most work of this kind involves inducing individuals or special groups to think, feel, and finally to act, wittingly or unwittingly, willingly or unwillingly, in a desired manner. To get them to act requires *inciting* them to act by so capitalizing on strong desires and emotions that they see and feel real personal or group advantages in acting, despite the personal danger involved, and *presenting* the lines of action in the proper way so they know clearly what should be done and find it within their means to do it. Getting people to act is not entirely a matter of intuition. A few broad principles are known to practitioners:

(1) Simplicity — Proposed ideas or actions should be simple and clear and be presented in a form easy to remember.

(2) Plausibility and feasibility—Proposed ideas or actions should be within the understanding of the persons or groups approached; and the desired work should be action they feel or are led to feel they can actually carry out.

(3) Suggestion (especially in deception) — Proposed ideas or actions are usually more effective when presented by indirection, so that persons or groups

SECRET

are led to believe they have arrived at their own conclusions, and that they voluntarily participate in the work.

e. INCITEMENT AND INDUCEMENT—Many persons or groups who will do effective MO work already have a strong desire for action. Verbal appeals to their own self-interests are likely to sound naive to them. They need little incitement — what they need is encouragement, guidance, and above all, support or subsidy. Fence-sitters, collaborationists, and enemy personnel with low morale can be induced to act subversively only if strongly impelled to do so in their immediate or future self-interest. It is necessary, therefore, to discover what will impel them to act, and, in terms of these impelling emotions and desires to use suggestion, bribery, or blackmail to induce action. In particular, the following should be exploited:

(1) Fears, anxieties, hatreds, feelings of discrimination, grievances, and mysticisms.

(2) Hopes and desires for personal gain.

(3) Conceptions of right and wrong.

(4) Social position (prestige, face, vanity).

(5) Desire for personal security.

30. *OPERATIONAL PLANS*

a. The MO Section Officer should delegate staff personnel to draw up detailed plans for every important MO tactical operation. These plans will show how the projected morale operations implement the approved special programs or directives based thereon.

b. These plans should be as detailed as possible, for they constitute a blue-print of action. They should be *written*, in order to insure that all important details have been considered, to facilitate conferring with collaborating personnel who are directly involved in the operations, to permit approval by the strategic services officer, and to provide for the briefing of operatives and agents who will put the plan into effect.

c. Responsibility for the development of the details of operational plans lies with the MO Section Officer.

SECRET

These plans will implement the MO Programs. The Implementation Studies sent from Washington will be useful in formulating these programs as well as specific information on which these plans are based obtained in the theater base from SI and R&A representatives, from G-2, A-2, ONI, JICA, and from Allied sources.

d. The nature of the problem will determine the form of an operational plan. It should, however, include the following:

(1) A precise delineation of the MO targets of the operation, with a brief statement indicating how the particular targets are, under the circumstances, the most suitable in implementing the special program.

(2) An adequate survey of intelligence bearing on the problem, including only such matters as collaborating personnel and MO personnel must know in order to execute the projected tasks.

(3) A complete detailing of all the important operations required. Progressive steps in the action should be indicated, showing the times when and places where action will occur, specific persons and groups involved, nature of cover, specific use of MO implements and the coordination of them, probable enemy counter-action to be anticipated, and how it is to be met.

(4) All details of collaboration with the armed services, with SO and SI, with Allied agencies, with native underground personnel, and with propaganda agencies (if involved).

e. To insure that all the operations are feasible, sufficient detail should be included showing the requirements of the operations and how they are met, covering such matters as equipment, communications, transport, supplies, liaison, and the selection and briefing of operatives and agents.

SECTION VII — OPERATIONS

31. *GENERAL*

a This section describes the subversive operations which may be especially suitable in carrying out the

SECRET

strategic tasks of MO as set forth in Section VI, paragraph 26 b.

b. Under each of these strategic tasks there is presented a statement of mission and of practical principles which may be helpful in planning and directing these operations.

c. This section does not attempt to show the detailed administration of personnel engaged in the subversive operations to accomplish these strategic tasks. In general, however, the strategic services officer will administer the personnel engaged in all morale operations including MO operatives and agents in enemy or enemy-occupied territory and at the base. These MO operatives and agents, though acting under the direction and policies of MO, will conduct their activities under the administration and discipline of an OSS officer responsible for operations of both MO and SO agents. Such activities will therefore call for the closest joint planning by the MO-SO staff personnel.

d. The actual implements and channels employed in these operations are also not developed in this section, except by way of illustration. The actual means available to the MO Section Officer in each theater base — and they may vary greatly from one base to another — will determine by what direct and indirect media he will effect these operations. At some bases, some of the implements required for MO work, especially radio and printing facilities, may be assigned by the theater commander to a special combat propaganda unit not under the jurisdiction of the strategic services officer. In this case, MO personnel, supplies, and equipment may be assigned to this unit to effect that part of the MO program requiring these implements. For effective action in this situation, the closest collaboration between the MO Section Officer and the officer in charge of this special unit is necessary.

32. *WITHIN THE ENEMY'S COUNTRY*

a. DIVIDING THE ENEMY

(1) The mission is to destroy unity among enemy groups and people by alienating group from group,

people from leaders, and nation from nation. A divided home front and a split between enemy nationals are major factors in enemy collapse.

(2) Divisive activities are most profitably carried on with greatest intensity following enemy military reverses or domestic crises, *since at these times dormant splits begin to reappear*. These divisive activities, whether on a local or on an international scale, concentrate on exploiting long-standing cleavages and issues.

(3) The following activities serve to divide *group from group*: launching subversive radio, press, or rumor programs which attack one group (e.g., the Army) but do not attack and may even defend another group (e.g., the Party); planting evidence of treason on political or military leaders by such means as: sending compromising decipherable messages by radio or telegraph, having the individual in question receive especially favorable mention in clandestine leaflets which will be picked up by the police; arranging leaks in neutral countries by diplomatic personnel revealing that selected individuals can be "depended upon"; spreading evidence to show that big industrialists are profiteering; spreading rumors that army divisions drawn from "politically difficult" areas are being sent to the most dangerous fronts; spreading rumors that more members of a given group or class have surrendered or deserted to Allied forces than others. Minor leaders of a political or economic group which has grievances against other groups should be encouraged by bribery and persuasion to take a militant stand against the offending group. The possibility of simultaneously encouraging and bribing the second group to similar aggressive action should also be investigated.

(4) Tactics that divide *people from leaders are*: spreading rumors that leaders are protecting their future by buying securities and durable goods in neutral countries; arranging for agents actually to purchase money, jewels or precious works of art in a neutral capital, then letting it leak that these are

being bought for prominent leaders; introducing into the enemy country picture postal cards of famous castles, villas, and in the descriptive paragraph including such sentences as "The recently purchased villa in Sweden of Herr von Ribbentrop is very similar in architectural design to this beautiful building"; spreading subversive songs, jokes, cartoons, which attack the government; sending tips to the secret police of alleged hoarding and black market operations; circulating reports that deplore or commend the brilliance of methods by which the sons and relatives of officials are avoiding the draft, getting easy jobs, buying restricted food.

(5) Where authentic evidence is available that some leaders are engaged in one or more of the illegal actions described above, opportunities should be put in their way for expanding these activities, to be followed by exposure or blackmail.

(6) To divide *nation from nation*, the following activities are effective; exploiting incidents in which friction occurs between Axis soldiers from different countries; provoking fights and brawls wherever different Axis nationals come together; circulating reports that the reason for moving prisoners of war from one camp to another is enmity between the nationals; circulating rumors and reports of all cases in which forces of the stronger Axis countries desert the troops of the weaker Axis members; circulating statistics of disproportionate casualties of the different nationals; when troops of one country are quartered in another Axis area, exploiting all social problems, especially sexual, which inevitably arise, increase of juvenile delinquency, adultery, pregnancy, venereal diseases; circulating in one enemy country political statements, jokes and satires derogatory to it which are ascribed to the nationals of the other.

b. INDUCING PANIC IN ENEMY POPULATIONS

(1) The mission is to augment feelings of insecurity, anxiety, and panic in enemy civilian population. Such a program is especially desirable just before a

SECRET

military offensive. But even when a battle is not imminent, neurotic anxiety adds greatly to the political and administrative problems of the enemy. Further, the mental state of the civilians at home inevitably is communicated to men at the front through letters or *via* soldiers returning from furloughs, and thus a panicky civilian population helps to demoralize soldiers in combat zones.

(2) Operations are preferred which augment normal fright reactions to war events, such as inducing sympathizers to scream hysterically in factories and shelters during air raids or upon the publication of casualty lists; distributing small clandestine leaflets purporting to tell how to avoid alleged horrible consequences of incessant bombings — insanity and feeblemindedness, stunting of growth, various nervous disorders, sexual impotency or frigidity, sterility, miscarriage, deafness, blindness; spreading "information" on the methods of avoiding these terrible effects, say, by recommending going to a doctor within 24 hours after a bombing for examination whether the person appears injured or not.

(3) One of the most disorganizing civilian anxieties is the belief that there exists a widespread fifth-column group, that there is no one in whose loyalty one can have faith. To exploit this fear is dangerous, as it may result in reprisals on bona fide agents and sympathizers. When, however, an hysterical state can be so fostered that mob action can be turned on civilians not in the underground, then this work is most effective.

c. STRENGTHENING THE ENEMY CIVILIANS' DESIRE FOR PEACE, AND RAISING FALSE HOPES

(1) The mission is to subvert the fighting spirit of enemy civilians by causing them to agitate for peace and by raising false hopes. Any degree of success not only adds to the crippling of enemy civilian war work, but communicates poor morale to the fighting forces.

(2) Enemy civilians who are susceptible to peace stimuli can be divided into two general categories:

SECRET

(a) Those who are resistant to the regime and who use pacifist propaganda as a cover to excuse their activities and protect them from charges of treason.

(b) Those who are loyal to the regime but sincerely desire peace.

(3) To exploit *both* these groups, clandestine agitation for peace should be kept on a lofty plane. It should not be complicated by tirades against enemy leaders. It should not be suspect as originating outside the country. Peace is urged as a desirable end in itself.

(4) Two major lines of action may be employed: playing up the horrors of war, and exploiting the pacifistic or religious attitudes of the civilians.

(5) For the horror approach, subversive pictures, for example, may be circulated of burned and mutilated soldiers, asking, "Will your son or husband be next? Save him from this by demanding peace." These pictures should be distributed ostensibly, say, by a peace society made of enemy underground nationals. The maimed and crippled may be encouraged to display themselves in public as much as possible. A rumor may be spread that a certain shrine in a large city has miraculously cured war injured. All war injured are encouraged to make pilgrimages to that shrine.

(6) Religious pacifism is exploited by suggestion; for example, numerous small crosses may be distributed with the single word "peace" written thereon. A rumor may be spread that because of the appearance of a certain saint who also appeared shortly after the end of the last war thousands of women are now praying for peace at noon every day (that being the hour designated by the saint). Reports may be circulated that certain women who have prayed for peace at a certain shrine (in a large city) have miraculously been spared death in the family.

(7) Desire and action for peace will be increased if enemy civilians are made to believe that there is real evidence that peace is within immediate reach. It is

SECRET

important that such "evidences" be confined to enemy territory. For example, reports may be circulated that engineers are already planning the conversion of enemy factories to peace-time production.

(8) When the object of circulating "evidence" of peace is the raising of false hopes, it is necessary that the hopes raised be "proved" in due course to be actually false. The natural course of events may provide such proof. However, it is best to plan for the pricking of the bubble whenever possible. This may be accomplished in two ways:

(a) By circulating the "evidences" of peace at an appropriate interval before the occurrence of an *actual event* (secretly known to OSS) which will prove the first reports of peace to be false.

(b) By following the first "evidences" of peace after an appropriate interval by "new evidence" which proves the first reports to be untrue.

d. SUBVERTING ENEMY CIVILIAN POPULATIONS DURING A GROUND OFFENSIVE

(1) The mission is two-fold: first, during offensive action in enemy home territory, to create such confusion and panic among civilians as will disrupt enemy military action and demoralize his troops, and second, to invite cooperation of war-weary civilians with our own invading forces so as to decrease our own military problems.

(2) Demoralization and terrorization attacks on civilian populations in enemy countries can, if not immediately followed through by military conquest, have the unintended effect of consolidating the group, improving its morale, and making it more determined than ever. Therefore, all-out terroristic operations should be carried out in areas that comprise the *immediate* objective of our armed forces.

(3) When the proper time and place have been determined, sympathizers and agents may be instructed to create false air raid alarms; start street fighting and riots; spread rumors that the town will

SECRET

be utterly destroyed by bombings; that Allied paratroopers have already landed in certain sections of the town; that enemy forces at one point have been utterly destroyed; that the enemy military (at times when their retreat seems imminent) have planted mines in all the city streets, have polluted the water supply.

(4) At this time a bribed police force is of particular value for augmenting or for not interfering with such disorders as may occur. Further, at this time the police force is particularly susceptible to bribery and subversion since the police force knows it must remain in the town after the enemy troops have evacuated it and will therefore be ready to be "persuaded" to change sides.

(5) A terrorized civilian group threatened with the devestation of military action will, unless provided with a feasible and hopeful means of escape and positive courses of action to implement it, develop a "back-to-the-wall" attitude, and become a dangerous consolidated adversary. All feasible subversive means of providing hope should therefore be employed. For example, reports of fair treatment which other enemy civilians have received at the hands of our forces may be circulated. Specific methods of signifying capitulation should be clandestinely circulated, such as staying indoors, displaying signs, flags, reporting hidden stores.

(6) Frequently, during ground operations directed against an inhabited point, terrorization can be employed to exploit the tendency of the population to flee from such towns or cities. It is possible, through the use of rumors both to speed up such pell-mell evacuation and to direct it along roads which the enemy would rather keep open and free from civilian traffic.

33. *WITHIN ENEMY - OCCUPIED OR CONTROLLED TERRITORY*

a. PROMOTION OF RESISTANCE AND REVOLT AGAINST THE ENEMY

SECRET

(1) The mission is to induce a greater resistance among the population and to promote the organization of the resistance under good leadership.

(2) OSS will not, in most cases, use its own operatives and agents in directly inciting potentially resistant masses of people. This job can usually best be done by selected native influential persons on the spot who know the people of the area. The OSS task is therefore reduced to supplying through personal contact and other inducements, materials, and guidance by means of which native persons can do the work of incitement.

(3) As for materials which MO personnel will prepare outside the territory (as by freedom stations), the following principles should be kept in mind:

(a) Direct and specific appeals should be made to the individual's fear of personal harm, social ostracism, prejudices, resentments, and hope for rewards.

(b) The impression should be created that the resistant movement is popular, that everyone is "joining up." Special signs, symbols, songs, etc., are issued.

(c) Since actual participation in a movement (*doing* something) tends to increase the feeling of personal identification with the cause of the movement, acts that *every* member of a resistant group can carry out should be urged (within limits of personal security). The simplest act, the smallest child should not be neglected. Simple instructional leaflets, or talks on freedom stations covering methods of simple sabotage may be effective.

(d) The important contribution of each act, especially the minor undetectable, subtle kinds of sabotage should be emphasized; how these acts divert troops, supplies, and officers; how they harass the enemy; how they slow down production or lift the morale of other resistant groups.

(e) The feeling of success increases the people's willingness to continue subversive work.

SECRET

Therefore, all clandestine (and cooperating white) means may be used to circulate reports and rumors of successful activities, especial care being taken, of course, to protect the people from reprisals.

(f) The personal safety of the citizen-saboteur should be safeguarded. Instruction should be given on methods of making sabotage appear to be accidental or the result of natural causes.

b. INTERFERING WITH THE ENEMY'S CONSOLIDATION AND USE OF AN OCCUPIED COUNTRY'S CAPABILITIES

(1) The mission is to hamper the enemy's efforts to exploit the industrial and civilian administration of an occupied country and to harass enemy government administration. Success in this work pins down enemy troops in troublesome areas and upsets enemy calculations concerning supplies based upon expected production. In addition, augmenting the "normal" problems of civilian administration will produce irritation, confusion and inefficiency in enemy or collaborationist administration.

(2) The technique most suitable for this work is that of fostering "simple sabotage" by the populace. Simple sabotage refers to inconspicuous acts which destroy enemy targets or obstruct enemy production, administration, and political consolidation. In contrast to major sabotage, simple sabotage is performed without the use of specially prepared tools or equipment; it is executed by an ordinary citizen who acts individually and without necessarily any active connection with an organized subversive group; and it is carried out in such a way as to involve a minimum danger of injury, detection and reprisal.

(3) Simple sabotage may consist of (a) *physical destruction* of equipment and installations, or (b) *functional interference* with operations. Detailed treatment of this subject will be covered in a Strategic Services Field Manual.

c. PRODUCING CIVILIAN DISORDER IN SUPPORT OF MILITARY OPERATIONS

(1) The mission is to deceive the enemy commander so that he may make false moves based upon

SECRET

incorrect estimates of civilian resistance. Inducing the enemy command erroneously to anticipate large civilian uprisings pins down enemy troops in such areas and assists our armed services if they wish such a diversion. On the other hand, successfully inducing the enemy command to believe that civilian spirit is low and that extra military precautions are unnecessary permits OSS and the armed services to operate with less enemy interference.

(2) Strategic and tactical planning at the highest military echelon is required in this work in order to assist the theater commander in his military program. The success of this work furthermore requires the closest collaboration with the underground.

(3) It is necessary, first, to determine whether to give the impression of spontaneous revolt or of an elaborately organized underground at work. The needs of the specific situation will provide the answer.

(4) If it is to be an impression of spontaneous and widespread revolt, minor subversive activities at widely scattered points are planned and numerous outbreaks committed simultaneously. Thus, at a predetermined time, civilians at various places break windows, start fires, street fights, and riots. Judicious and effective bribing of police officials can help in the successful staging of street fights and riots, without endangering the security of the agent. The police can be bribed either to stay away from the scene until it has developed into a riot of its own momentum, or to make wholesale arrests of innocent by-standers and thus create the impression of greater participation in the riot than actually existed.

(5) If it is to be the impression of an elaborately organized underground at work, coordinated sabotage acts aimed at destroying a single extensive facility are planned. Thus, sympathizers are incited to carry out, under clandestine direction, certain sabotage acts, all of which when taken together tend to demolish or disrupt a given communication, transport or supply system.

(6) Where the enemy is to experience a false sense

SECRET

of security, sympathizers are clandestinely instructed to cease from all kinds of sabotage. In accordance with the principle of proper timing, it is important that, when possible, such instructions are given after a particularly determined attempt of the enemy authorities to prevent sabotage. The sudden cessation of sabotage activities at any other time would tend to put the enemy authorities even more on their guard.

d. INCREASING TERROR, FRICTION, AND DEMORALIZATION AMONG COLLABORATIONISTS

(1) The mission is to create friction between collaborationists and Axis officials, and among collaborationists themselves, and to incite sympathizers to engage in terroristic operations against collaborationists. Disruption of the work of collaborationists increases the difficulties of the enemy command inasmuch as the collaborationists relieve the enemy military command of many of the responsibilities of civilian administration.

(2) The period during major battles is a period for increased terroristic acts of this sort. Extensive and all-out terroristic activities should be urged through every available channel and individual initiative in these activities is encouraged against bona fide collaborationists. Such acts as the following are suitable methods of terrorizing collaborationists: inciting the populace to dispose of them and, where possible, to plant evidence at the scene indicating the existence of a powerful society which has set itself up to deal with all collaborationists; spreading black lists of quislings; smearing symbols on their houses and vehicles and committing terroristic acts against them; writing anonymous letters and making anonymous telephone calls threatening the lives of collaborationists and their families; preparing documents to be "discovered" by the collaborationists which will appear to give directions for the disposition of collaborationists on D-day; deluging religious authorities with such theological questions as whether or not a quisling who is hanged can be buried in consecrated ground.

SECRET

(3) Projects designed to create friction among collaborationists should be geared into combat activities wherever possible. Most effective are acts which suggest to the Axis officials that the collaborationist is "selling out" or is untrustworthy in a battle situation. For example, "secret" documents may be discovered by the military officials listing the names of collaborationists who have allegedly appealed to the underground for clemency and who have affirmed their readiness to come over to our side on D-day. Quislings may be tipped off with false information which they transmit to enemy authorities. Tips may be sent to enemy officials and secret police concerning collaborationists who are dealing with the enemy, with underground forces.

(4) Collaborationists, both political and industrial, are concerned with questions of power and status. This battle for status may be exploited by such means as preparing a document ostensibly originating with one group of collaborationists protesting against the activities of another group of collaborationists, who "find" the document.

e. PROVOKING REBELLION OR COUP D'ETAT IN A SATELLITE COUNTRY OR INDUCING ITS SEPARATION FROM THE AXIS

(1) The mission is to aid in the inciting and carrying through of revolutions, incidents, changes in governments, or coup d'etat in satellite or other enemy-dominated countries. Since many of the operations are in the nature of secret diplomacy, close consultation with the State Department may be necessary.

(2) In many cases, the contribution of OSS will consist of supplying the State Department or our military commanders with channels of communication or with agents who can conduct necessary secret negotiations. Frequently, however, through its contacts with the underground or through neutral countries, OSS may discover and suggest to the State Department opportunities for political or economic operations against the enemy. Influential persons in

SECRET

a satellite country who are in a strong enough position to evoke a coup d'etat, rebellion, or other important strategic act against a collaborationist regime will often require postwar commitments. OSS is not authorized to make such commitments. These must come from the highest authority.

(3) A second category of activities is that of provoking an "incident" between the enemy country and its satellites, and then suitably exploiting the "incident" by freedom stations and false leaflets. In the provoking of an "incident" the *agent provocateur* usually plays an important role. For example, such an agent who has succeeded in establishing himself as an enemy official can be tactlessly or ruthlessly carrying out administrative orders—increase the antagonism of the people of the country against the enemy government. The agent who has penetrated a pro-Axis political group in a satellite country can incite other members to militant action or to ill-considered policies and thus help create incidents which will discredit the pro-Axis group, or which will call for retaliatory action from opposition groups.

34. *WITHIN OTHER AREAS*

a. Establishing an MO Section of a Base in a Neutral Country for Operations in Enemy and Enemy-Dominated Countries

(1) All SS activities must be under the direction of the Chief of OSS mission. Such activities may be in support of programs initiated in other theaters or areas or may be activities pursuant to another approved program.

(2) The mission is to make the necessary personal contacts and to utilize neutral implements and channels for such operations. Establishing a neutral base for these purposes is a critical necessity, for many of the missions described in other parts of this section require operating from a neutral base.

(3) No MO activity should result in the jeopardy of national policy, as expressed and put into effect by the chief of the diplomatic mission. Hence, he should

SECRET

be apprised in general, of MO activities in the area to which he is assigned.

(4) Personal contact work will usually be of three sorts:

(a) Contacts through appropriate channels with influential persons in enemy-dominated countries and, when the situation within the enemy country itself promises internal collapse with influential persons therein.

(b) Contacts with the underground that is collaborating with SS for MO work in areas bordering the neutral country.

(c) Such contacts as will facilitate the introduction into bordering areas of SS agents engaged in special MO missions.

(5) Direct use of neutral implements and channels for MO purposes may often be possible. For example, rumors designed for consumption in enemy areas may be spread by word of mouth or by newspaper or radio "plants." Forged documents may be planted in such a way as to fall in enemy hands. The mail into enemy zones may be used for transmitting poison-pen or other subversive materials.

(6) The special status of a neutral country requires the closest collaboration between SS personnel doing MO work and other SS personnel. In the first place, the enemy has legitimate access to and representatives in the country. There is, therefore, continual danger of enemy penetration into the SS organization. MO personnel should work closely, if indirectly, with SS men assigned the duty of protecting OSS from penetration, and should individually and vigilantly observe security measures.

(7) In the second place, SS intelligence men in a neutral country are in a special position to provide information on MO targets of opportunity. This advantage comes from the presence of official representatives of the enemy, refugees, neutral observers, and enemy nationals travelling into and out of enemy

SECRET

territory. When operations are planned for such targets of opportunity, care should be exercised to see that they are consistent with the authorized over-all MO program.

b. ASSISTING THE CHIEF OF THE DIPLOMATIC MISSION IN SPECIAL WORK REQUESTED BY HIM

The mission is to assist the chief of the diplomatic mission by conducting any operations requested by him that fall within the province and capabilities of MO. The enemy will usually conduct subversive activities in the neutral country, either openly, or by a fifth column, or by native persons or organizations which collaborate with him. In order to protect the United States war policy and interests, the chief of diplomatic missions may find it necessary to call upon OSS for special MO work deemed by him not to be expressly prejudicial to the relations between the United States and the neutral country. Any such work should be authorized by OSS, Washington.

35. *WITHIN THE ENEMY ARMED FORCES, IN ALL AREAS*

a. FOSTERING REBELLIOUSNESS WITHIN ENEMY ARMED FORCES

(1) The mission is to increase in enemy troops a spirit of rebelliousness against their officers and political leadership, especially among occupation forces. Even partial achievement of this aim would tend to communicate the rebellious spirit to soldiers in combat areas.

(2) Rebelliousness among soldiers occurs when they lose fighting spirit, and develop acute friction among themselves.

(3) To weaken the fighting spirit, advantage is taken of the fact that the enemy soldier, stationed behind the lines, has most of his contacts with civilians and tends to think, hope, and fear very much like a civilian. Most subversive work, therefore, focuses on the *civilian* attitudes of these soldiers. Every attempt is made to induce anxiety among the troops about conditions at home and to strengthen their

nostalgia for home. For example, rumors may be circulated about the conditions of families due to air raids, disease, overwork, adultery with foreign workers or with stay-at-home soldiers or civilians; letters from foreign workers in the home country may also be forged, describing the above conditions, and then allowed to fall into the hands of the enemy troops. To increase the feeling of isolation, civilians may be urged to talk sympathetically with enemy soldiers about the dangers at home confronting their children, wives, and relatives, but in marked contrast, to treat them coldly during their own church festivals, family gatherings.

(4) To increase friction and mutual distrust among enemy military personnel, *natural* cleavages based on differences in status and ethics among the troops are the logical points of attack. Such cleavages may be discovered between enlisted men and their officers; between soldiers from one nationality or geographic area and soldiers from another nationality and area; or between one group of soldiers with one set of traditions and political ideology and another group of soldiers with another set (as between the Wehrmacht and the SS). Thus, wherever political police of other classes or soldiers have special privileges, clandestine leaflets may be distributed describing these privileges in detail. Ridicule is employed, e.g., toilets are marked "Reserved for Gestapo." Poison-pen letters may be written to officers and men implicating others who are disliked in treasonable, immoral, or unsporting acts. Rumors may be spread indicating how the political police are growing fat on corruption, are at home enjoying the wives and sweethearts of the front-line soldiers, how the political and military elite are preparing for escape to neutral countries in anticipation of losing the war. Attempts may be made to cause some troops to feel that they are sharing greater risks than others, or that some will incur reprisals for "war crimes" while others will not be held responsible.

b. INDUCING SURRENDER

(1) The mission is to increase the demoralization of enemy troops to the point of greater willingness to surrender. The voluntary surrendering of enemy soldiers obviously increases the relative strength of our armed forces, and reduces the fighting spirit of the remaining enemy troops.

(2) This work can be most effectively carried out only when the morale of enemy soldiers is low, especially during and after military reverses, or when they are green and untrained.

(3) The enemy soldier fears annihilation on the one hand, and an imagined harsh treatment as prisoner on the other. The contrast between horrible destruction and decent treatment as prisoner is therefore plausibly built up through subversive means. For example, rumors may be floated of the number of Axis soldiers who have broken down mentally under terrific air attack or cannon barrage. Sympathizers may be urged to contaminate enemy food, and burn enemy buildings and supplies during confused enemy military operations. Timed with these activities, rumors, reports and evidences of good treatment of prisoners should be circulated by every means behind enemy lines.

(4) On grounds of military ethics, the enemy soldier believes that surrender gives him dishonorable status as a soldier. He must be made to believe that this is untrue. Reports may therefore be circulated about wholesale desertion and surrender at other sectors, about "respect" accorded surrender, e.g., officers may retain their sidearms.

(5) The soldier fears for his security in the act of surrendering; he may be shot by his officers or by our own forces. He must be informed how not to reveal to his officers his intent to surrender, and learn to look for leaflets, pass words, gestures, which will assure safe passage.

(6) Such morale subversion activities should be timed with and support open front propaganda.

SECRET

Though often effective, official front appeals by radio, leaflets are too obviously a "selling-job." Such appeals "substantiated" by allegedly bona fide reports and rumors within and behind enemy lines carry greater weight.

36. *WITHIN OSS, WASHINGTON AND THEATERS*

a. WITHIN OSS, WASHINGTON, OVER-ALL PLANNING FOR AND SERVICING OF FIELD OPERATIONS

For a description of the planning and servicing activities of OSS, Washington, see Section II.

b. ASSISTANCE DESIRED BY THEATER COMMANDER

The theater commander may request to use MO personnel and equipment for special combat subversive activities in forward areas in support of United States Army operations. OSS personnel may, for example, be asked to serve as members of intelligence squads having as a special mission the collection of information useful to subversive work, for example, the locating and appraisal of radio, press and other equipment in forward enemy areas. OSS may also be asked to contribute plans, linguists, technical help and equipment to operating crews engaged in combat subversion or deception, including the operating of mobile radios, loud speakers and presses.

www.ingramcontent.com/pod-product-compliance
Lightning Source LLC
Chambersburg PA
CBHW050240230526
45470CB00005B/2041